ENERGY SECTOR STANDARD OF THE PEOPLE'S REPUBLIC OF CHINA

中华人民共和国能源行业标准

Code for Design of Underground Steel Bifurcated Pipe with Crescent Rib of Hydropower Stations

水电站地下埋藏式月牙肋钢岔管设计规范

NB/T 35110-2018

Chief Development Department: China Renewable Energy Engineering Institute

Approval Department: National Energy Administration of the People's Republic of China

Implementation Date: July 1, 2018

China Water & Power Press

中国水利水电出版社

Beijing 2024

All rights reserved. No part of this publication may be reproduced, stored in a retrieval system, or transmitted in any form or by any means—electronic, mechanical, photocopying, recording or otherwise, without prior written permission of the publisher.

图书在版编目（CIP）数据

水电站地下埋藏式月牙肋钢岔管设计规范：NB/T 35110-2018 = Code for Design of Underground Steel Bifurcated Pipe with Crescent Rib of Hydropower Stations (NB/T 35110-2018) : 英文 / 国家能源局发布. -- 北京 : 中国水利水电出版社, 2024. 9. -- ISBN 978-7-5226-2737-3

Ⅰ. TV732-65

中国国家版本馆CIP数据核字第2024LW9274号

ENERGY SECTOR STANDARD
OF THE PEOPLE'S REPUBLIC OF CHINA
中华人民共和国能源行业标准

Code for Design of Underground Steel Bifurcated Pipe
with Crescent Rib of Hydropower Stations
水电站地下埋藏式月牙肋钢岔管设计规范
NB/T 35110-2018
（英文版）

Issued by National Energy Administration of the People's Republic of China
国家能源局　发布
Translation organized by China Renewable Energy Engineering Institute
水电水利规划设计总院　组织翻译
Published by China Water & Power Press
中国水利水电出版社　出版发行
　　Tel: (+ 86 10) 68545888　68545874
　　sales@mwr.gov.cn
　　Account name: China Water & Power Press
　　Address: No.1, Yuyuantan Nanlu, Haidian District, Beijing 100038, China
　　http://www.waterpub.com.cn
中国水利水电出版社微机排版中心　排版
北京中献拓方科技发展有限公司　印刷
184mm×260mm　16开本　4.25印张　134千字
2024年9月第1版　2024年9月第1次印刷
Price（定价）：￥700.00

Introduction

This English version is one of China's energy sector standard series in English. Its translation was organized by China Renewable Energy Engineering Institute authorized by National Energy Administration of the People's Republic of China in compliance with relevant procedures and stipulations. This English version was issued by National Energy Administration of the People's Republic of China in Announcement [2023] No. 1 dated February 6, 2023.

This version was translated from the Chinese Standard NB/T 35110-2018, *Code for Design of Underground Steel Bifurcated Pipe with Crescent Rib of Hydropower Stations*, published by China Water & Power Press. The copyright is reserved by National Energy Administration of the People's Republic of China. In the event of any discrepancy in the implementation, the Chinese version shall prevail.

Many thanks go to the staff from the relevant standard development organizations and those who have provided generous assistance in the translation and review process.

For further improvement of the English version, any comments and suggestions are welcome and should be addressed to:

China Renewable Energy Engineering Institute
No. 2 Beixiaojie, Liupukang, Xicheng District, Beijing 100120, China
Website: www.creei.cn

Translating organization:

POWERCHINA Beijing Engineering Corporation Limited

Translating staff:

GAO Yan   QI Wen   GUO Jie   CHE Zhenying

Review panel members:

JIN Feng	Tsinghua University
LIU Xiaofen	POWERCHINA Zhongnan Engineering Corporation Limited
QIE Chunsheng	Senior English Translator
YAN Wenjun	Army Academy of Armored Forces, PLA
CHEN Lei	POWERCHINA Zhongnan Engineering Corporation Limited

JIA Haibo   POWERCHINA Kunming Engineering Corporation Limited

XU Xiaorong   North China Electric Power University

National Energy Administration of the People's Republic of China

翻译出版说明

本译本为国家能源局委托水电水利规划设计总院按照有关程序和规定，统一组织翻译的能源行业标准英文版系列译本之一。2023 年 2 月 6 日，国家能源局以 2023 年第 1 号公告予以公布。

本译本是根据中国水利水电出版社出版的《水电站地下埋藏式月牙肋钢岔管设计规范》NB/T 35110—2018 翻译的，著作权归国家能源局所有。在使用过程中，如出现异议，以中文版为准。

本译本在翻译和审核过程中，本标准编制单位及编制组有关成员给予了积极协助。

为不断提高本译本的质量，欢迎使用者提出意见和建议，并反馈给水电水利规划设计总院。

地址：北京市西城区六铺炕北小街 2 号
邮编：100120
网址：www.creei.cn

本译本翻译单位： 中国电建集团北京勘测设计研究院有限公司
本译本翻译人员： 高 燕 齐 文 郭 洁 车振英
本译本审核人员：

金 峰 清华大学

刘小芬 中国电建集团中南勘测设计研究院有限公司

郄春生 英语高级翻译

闫文军 中国人民解放军陆军装甲兵学院

陈 蕾 中国电建集团中南勘测设计研究院有限公司

贾海波 中国电建集团昆明勘测设计研究院有限公司

徐小蓉 华北电力大学

国家能源局

Announcement of National Energy Administration of the People's Republic of China [2018] No. 4

According to the requirements of Document GNJKJ [2009] No. 52, "Notice on Releasing the Energy Sector Standardization Administration Regulations (*tentative*) and detailed implementation rules issued by National Energy Administration of the People's Republic of China", 168 sector standards such as *Guide for Evaluation of Vibration Condition for Wind Turbines*, including 56 energy standards (NB) and 112 electric power standards (DL), are issued by National Energy Administration of the People's Republic of China after due review and approval.

Attachment: Directory of Sector Standards

National Energy Administration of the People's Republic of China

April 3, 2018

Attachment:

Directory of Sector Standards

Serial number	Standard No.	Title	Replaced standard No.	Adopted international standard No.	Approval date	Implementation date
...						
30	NB/T 35110-2018	Code for Design of Underground Steel Bifurcated Pipe with Crescent Rib of Hydropower Stations			2018-04-03	2018-07-01
...						

Foreword

According to the requirements of Document GNKJ [2014] No. 298 issued by National Energy Administration of the People's Republic of China, "Notice on Releasing the Development and Revision Plan of the First Batch of Energy Sector Standards in 2014", and after extensive investigation and research, summarization of practical experience, and wide solicitation of opinions, the drafting group has prepared this code.

The main technical contents of this code include: layout principles and patterns, materials, basic design requirements, selection of shape parameters, structural design, detailing requirements, testing, and safety monitoring.

National Energy Administration of the People's Republic of China is in charge of the administration of this code. China Renewable Energy Engineering Institute has proposed this code and is responsible for its routine management. Energy Sector Standardization Technical Committee on Hydropower Investigation and Design is responsible for the explanation of specific technical contents. Comments and suggestions in the implementation of this code should be addressed to:

China Renewable Energy Engineering Institute
No. 2 Beixiaojie, Liupukang, Xicheng District, Beijing 100120, China

Chief development organization:

POWERCHINA Beijing Engineering Corporation Limited

Participating development organizations:

POWERCHINA Zhongnan Engineering Corporation Limited

POWERCHINA Huadong Engineering Corporation Limited

POWERCHINA Guiyang Engineering Corporation Limited

POWERCHINA Kunming Engineering Corporation Limited

Chief drafting staff:

WANG Zhiguo	LYU Mingzhi	LI Zhenzhong	WANG Jianhua
GENG Guibiao	GAO Yonghui	JIANG Kuichao	GUAN Lihai
QIAN Yuying	SONG Ruixiang	WANG Yonghui	HU Wangxing
CHEN Lifen	LUO Yuxia	LIU Xiangmin	

Review panel members:

DANG Lincai	FANG Guangda	HAO Jungang	SUN Yirong
LU Qiang	WU Hegao	WU Wenping	XIONG Chungeng
CHEN Zihai	YANG Xingyi	YANG Lina	FENG Shineng
YAO Minjie	YANG Xiaomei	ZHENG Linping	LI Shisheng

Contents

1	General Provisions	1
2	Terms	2
3	Layout Principle and Patterns	4
4	Materials	5
5	Basic Design Requirements	16
6	Selection of Shape Parameters	22
7	Structural Design	24
8	Detailing Requirements	27
9	Testing	31
9.1	Model Test	31
9.2	Hydrostatic Test	31
10	Safety Monitoring	34
Appendix A	Shape Design Methods for Underground Crescent-Rib Reinforced Bifurcations	37
Appendix B	Structural Analysis Methods for Underground Crescent-Rib Reinforced Bifurcations	47
Explanation of Wording in This Code		55
List of Quoted Standards		56

1　General Provisions

1.0.1　This code is formulated with a view to standardizing the design of underground steel bifurcated pipe with crescent rib of hydropower stations, to ensure the design quality and achieve operational safety, cost effectiveness and technological advancement.

1.0.2　This code is applicable to the design of underground crescent-rib reinforced bifurcations for Grades 1, 2 and 3 water conveyance systems of hydropower stations, with a main pipe diameter no more than 9.0 m and a steel bifurcation HD value no greater than 4500 m · m. For steel bifurcations exceeding the above range, a special study and demonstration shall be conducted.

1.0.3　The design of underground crescent-rib reinforced bifurcations shall, according to the practical situations of the project, select reasonable layout, materials, structural type, detailing and measures, to meet the requirements for bearing capacity, stability, stiffness, and corrosion resistance during fabrication, transportation, installation, hydrostatic test and operation.

1.0.4　In addition to this code, the design of underground steel bifurcated pipe with crescent rib of hydropower stations shall comply with other current relevant standards of China.

2 Terms

2.0.1 steel bifurcation

section of penstock at bifurcating position, consisting of the main body of the bifurcation and part of the main pipe and branch pipes

2.0.2 crescent-rib reinforced bifurcation

steel bifurcation internally reinforced with a crescent plate at the bifurcating position

2.0.3 exposed pipe state

stress state of the steel bifurcation bearing internal water pressure solely

2.0.4 buried pipe state

stress state of the steel bifurcation bearing internal water pressure jointly with surrounding rocks

2.0.5 critical external buckling pressure

maximum pressure calculated in the design of steel bifurcations, under which the steel bifurcation can withstand external pressure and keep stable

2.0.6 general membrane stress

stress generated under internal water pressure and well distributed along the thickness, satisfying the simple laws of equilibrium, with the influencing range over the whole structure and without self-limitation

2.0.7 local membrane stress

stress generated under internal water pressure and well distributed along the thickness, satisfying the deformable compatibility conditions resulting from the whole structure discontinuity attributable to the discontinuity of generatrix at the connection of different cones of pipe shell, with the influencing range limited to structural local areas

2.0.8 bending stress

normal stress varying along pipe wall thickness, whose maximum value occurs on the surface of pipe wall

2.0.9 bifurcation scale

expressed by HD value, where H is the design internal water pressure of the steel bifurcation (m) and D is the diameter of the main pipe (m)

2.0.10 half-apex angle

angle between the axis of the cone and a generatrix

2.0.11 common-tangent sphere

sphere tangent to adjacent cone pipes

2.0.12 amplification ratio

ratio of the radius of the maximum common-tangent sphere to the radius of adjacent main pipe

2.0.13 rib-width ratio

ratio of the width of the rib plate waist section to the horizontal projection length of the intersecting line between the rib plate and the middle surface of the pipe shell, where the width of the rib plate waist section is the width between the inner edge of the rib plate and the intersecting line between the rib plate and the middle surface of the pipe shell

2.0.14 waistline

intersecting line between the horizontal symmetrical plane of the steel bifurcation and the outer edge of the steel bifurcation shell

2.0.15 turning angle

deflection angle of the waistline of the adjacent pipe segments

2.0.16 average sharing ratio of surrounding rock

ratio of the difference between the mean Mises stress in steel bifurcation shell in exposed pipe state and that in buried pipe state to the mean Mises stress in exposed pipe state

2.0.17 exposed bifurcation criterion

criterion that either the sum of the local membrane stress and the bending stress in the steel bifurcation or the maximum stress in the rib plate calculated according to the normal operating conditions is not greater than the yield strength of the steel, without considering the load sharing by surrounding rocks

3 Layout Principle and Patterns

3.0.1 The layout of underground crescent-rib reinforced bifurcations shall comply with the current sector standard NB/T 35056, *Design Code for Steel Penstocks of Hydroelectric Stations*.

3.0.2 In order to reasonably control the steel bifurcation scale, the diameter of the main pipe of the steel bifurcation may be appropriately reduced through a techno-economic comparison, and the elevation of steel bifurcation may be raised appropriately when necessary.

3.0.3 When the *HD* value of steel bifurcations is not less than 2500 m·m, symmetrical Y-shaped layout should be adopted or the degree of asymmetry of steel bifurcations be minimized.

3.0.4 The bifurcating angle of symmetrical Y-shaped crescent-rib reinforced bifurcation should be the same as that of the pipeline axis; the bifurcating angle of asymmetrical Y-shaped crescent-rib reinforced bifurcation may be determined according to the layout conditions and the structural and hydraulic characteristics of the steel bifurcation and may be different from that of the pipeline axis.

4 Materials

4.0.1 The steels used for steel bifurcations shall be selected according to the steel bifurcation scale, service conditions, steel properties, fabrication and installation processes, cost effectiveness, etc. Steel bifurcations should be made of the steels of the grades listed in the current standards of China GB/T 700, *Carbon Structural Steels*; GB/T 1591, *High Strength Low Alloy Structural Steels*; GB/T 31946, *Steel Plates for Steel Penstock in Hydropower Station*; GB/T 713, *Steel Plates for Boilers and Pressure Vessels*; and YB/T 4137, *Low Welding Crack Susceptibility for High Strength Steel Plates*, or may be made of the steels of the grades listed in the current national standards GB/T 16270, *High Strength Structural Steel Plates in the Quenched and Tempered Condition*; and GB/T 19189, *Quenched and Tempered High Strength Steel Plates for Pressure Vessels*, the quality of which shall not be inferior to Class C.

4.0.2 When other steels than listed in this code or foreign standard compliant steels are used, their chemical composition, mechanical properties, and welding performance shall not be inferior to those of equivalent grades specified in the standards mentioned in Article 4.0.1 of this code.

4.0.3 The guarantees for steels used for steel bifurcations shall be specified as follows:

1   Cold bending test shall be carried out.

2   Weldability and toughness of welded joints shall be guaranteed, and the welding materials used shall match the base metal. The strength of welded joints shall not be lower than the characteristic value of the strength of the base metal.

3   Impact toughness indexes, impact test temperatures, sampling positions, etc. shall comply with the current applicable standards of China, and the sampling direction shall be transverse. Supplementary requirements may also be proposed depending on the project-specific operating conditions.

4   Each project may set requirements for the strain-aging sensitivity coefficient or the impact energy absorption after strain aging for steels according to the specific operating conditions.

5   The steels for the rib plates of the steel bifurcations shall also comply with the current national standard GB/T 5313, *Steel Plates with Through-Thickness Characteristics*, and each of original rolled steel

plates shall be inspected. The Z-direction performance level of the rib plates of crescent-rib reinforced bifurcation may be selected in accordance with Table 4.0.3.

Table 4.0.3 Z-direction performance level of rib plates of crescent-rib reinforced bifurcation

Plate thickness (mm)	Z-direction performance level
$t < 35$	–
$35 \leq t < 70$	Z15
$70 \leq t < 110$	Z25
$110 \leq t < 150$	Z35

6 All steel plates used for steel bifurcations shall be subjected to ultrasonic testing in accordance with the current national standard GB/T 2970, *Method for Ultrasonic Testing of Thicker Steel Plates*. The carbon steel and low-alloy steel shall conform to Grade Ⅲ and high-strength steel shall conform to Grade Ⅱ. The low-carbon steel, low-alloy steel and high-strength steel for crescent-rib reinforced bifurcations stressed in thickness direction shall conform to Grade Ⅰ.

4.0.4 The characteristic values and design values of strength for steel plates of underground steel bifurcations shall be selected in accordance with Table 4.0.4. When the ratio of yield strength to the tensile strength is not greater than 0.7, the characteristic value of strength f_{sk} in the table shall be R_e; otherwise, f_{sk} shall be $0.7R_m$.

4.0.5 The elastic modulus E_s of steels may be taken as $2.06 \times 10^5 \text{ N/mm}^2$, Poisson's ratio v_s as 0.3, linear expansion coefficient α_s as $1.2 \times 10^{-5}/\text{°C}$, and gravity γ_s as $78.5 \times 10^{-6} \text{ N/mm}^3$.

Table 4.0.4 Characteristic values and design values of strength for steel plates for underground steel bifurcations

Steel type	Grade (Applicable standard)	Delivery state	Steel plate thickness (mm)	Strength at normal temperature		Tensile, compressive, bending strength	
				Yield strength R_e (N/mm^2)	Tensile strength R_m (N/mm^2)	Characteristic value f_{sk} (N/mm^2)	Design value f_s (N/mm^2)
Carbon structural steel	Q235 (GB/T 700)	Hot rolled, control rolled or normalized	≤16	235	370	235	215
			>16 - 40	225		225	205
			>40 - 60	215		215	200
			>60 - 100	215		215	200
			>100 - 150	195		195	180
	Q275 (GB/T 700)		≤16	275	410	275	245
			>16 - 40	265		265	240
			>40 - 60	255		255	230
			>60 - 100	245		245	220
			>100 - 150	225		225	200

Table 4.0.4 (continued)

Steel type	Grade (Applicable standard)	Delivery state	Steel plate thickness (mm)	Strength at normal temperature		Tensile, compressive, bending strength	
				Yield strength R_e (N/mm^2)	Tensile strength R_m (N/mm^2)	Characteristic value f_{sk} (N/mm^2)	Design value f_s (N/mm^2)
Low-alloy high-strength structural steel	Q345 (GB/T 1591)	Hot rolled, control rolled, normalized, normalized rolled or normalized plus tempered, thermomechanical rolled (TMCP) or thermomechanical rolled plus tempered	≤16	345	470	330	295
			>16 - 40	335	470	330	295
			>40 - 63	325	470	325	290
			>63 - 80	315	470	315	280
			>80 - 100	305	470	305	275
			>100 - 150	285	450	285	305
	Q390 (GB/T 1591)		≤16	390	490	340	305
			>16 - 40	370	490	340	305
			>40 - 63	350	490	340	305
			>63 - 80	330	490	330	295
			>80 - 100	330	490	330	295
			>100 - 150	310	470	310	280

Table 4.0.4 (continued)

Steel type	Grade (Applicable standard)	Delivery state	Steel plate thickness (mm)	Strength at normal temperature		Tensile, compressive, bending strength	
				Yield strength R_e (N/mm^2)	Tensile strength R_m (N/mm^2)	Characteristic value f_{sk} (N/mm^2)	Design value f_s (N/mm^2)
Low-alloy high-strength structural steel	Q420 (GB/T 1591)	Hot rolled, control rolled, normalized, normalized rolled or normalized plus tempered, thermomechanical rolled (TMCP) or thermomechanical rolled plus tempered	≤ 16	420	520	365	325
			> 16 - 40	400	520	365	325
			> 40 - 63	380	520	365	325
			> 63 - 80	360	520	360	325
			> 80 - 100	360	520	360	325
			> 100 - 150	340	500	340	305
	Q460 (GB/T 1591)		≤ 16	460	550	385	345
			> 16 - 40	440	550	385	345
			> 40 - 63	420	550	385	345
			> 63 - 80	400	550	385	345
			> 80 - 100	400	550	385	345
			> 100 - 150	380	530	370	330

Table 4.0.4 (continued)

Steel type	Grade (Applicable standard)	Delivery state	Steel plate thickness (mm)	Strength at normal temperature			Tensile, compressive, bending strength	
				Yield strength R_e (N/mm²)	Tensile strength R_m (N/mm²)		Characteristic value f_{sk} (N/mm²)	Design value f_s (N/mm²)
Low-alloy high-strength structural steel	Q500 (GB/T 1591)	Hot rolled, control rolled, normalized, normalized rolled or normalized plus tempered, thermomechanical rolled (TMCP) or thermomechanical rolled plus tempered	≤16	500	610		425	380
			>16 - 40	480	610		425	380
			>40 - 63	470	600		420	375
			>63 - 80	450	590		410	370
			>80 - 100	440	540		375	340
	Q550 (GB/T 1591)		≤16	550	670		470	420
			>16 - 40	530	670		470	420
			>40 - 63	520	620		430	390
			>63 - 80	500	600		420	375
			>80 - 100	490	590		410	370

Table 4.0.4 (continued)

Steel type	Grade (Applicable standard)	Delivery state	Steel plate thickness (mm)	Strength at normal temperature		Tensile, compressive, bending strength	
				Yield strength R_e (N/mm^2)	Tensile strength R_m (N/mm^2)	Characteristic value f_{sk} (N/mm^2)	Design value f_s (N/mm^2)
Low-alloy high-strength structural steel	Q620 (GB/T 1591)	Hot rolled, control rolled, normalized, normalized rolled or normalized plus tempered, thermomechanical rolled (TMCP) or thermomechanical rolled plus tempered	≤ 16	620	710	495	445
			> 16 - 40	600	710	495	445
			> 40 - 63	590	690	480	435
			> 63 - 80	570	670	470	420
	Q690 (GB/T 1591)		≤ 16	690	770	540	485
			> 16 - 40	670	770	540	485
			> 40 - 63	660	750	525	470
			> 63 - 80	640	730	510	460
Quenched and tempered steel plate for high-strength structure	Q460 (GB/T 16270)	Quenched and tempered	≤ 50	460	550	385	345
			50 - 100	440	550	385	345
			> 100 - 150	400	500	350	315

Table 4.0.4 *(continued)*

Steel type	Grade (Applicable standard)	Delivery state	Steel plate thickness (mm)	Strength at normal temperature		Tensile, compressive, bending strength	
				Yield strength R_e (N/mm^2)	Tensile strength R_m (N/mm^2)	Characteristic value f_{sk} (N/mm^2)	Design value f_s (N/mm^2)
Quenched and tempered steel plate for high-strength structure	Q500 (GB/T 16270)	Quenched and tempered	≤ 50	500	590	410	370
			50 - 100	480	590	410	370
			> 100 - 150	440	540	375	340
	Q550 (GB/T 16270)		≤ 50	550	640	445	400
			50 - 100	530	640	445	400
			> 100 - 150	490	590	410	370
	Q620 (GB/T 16270)		≤ 50	620	700	490	440
			50 - 100	580	700	490	440
			> 100 - 150	560	650	455	410
	Q690 (GB/T 16270)		≤ 50	690	770	540	485
			50 - 100	650	760	530	480
			> 100 - 150	630	710	495	445

Table 4.0.4 *(continued)*

Steel type	Grade (Applicable standard)	Delivery state	Steel plate thickness (mm)	Strength at normal temperature		Tensile, compressive, bending strength	
				Yield strength R_e (N/mm^2)	Tensile strength R_m (N/mm^2)	Characteristic value f_{sk} (N/mm^2)	Design value f_s (N/mm^2)
Quenched and tempered high-strength steel plate for pressure vessel	07MnMoVR 07MnNiVDR 07MnNiMoDR (GB/T 19189)	Quenched and tempered	10 - 60	490	610	425	380
Steel plate for boiler and pressure vessel	Q245R (GB/T 713)	Hot rolled, control rolled or normalized	3 - 16	245	400	245	220
			> 16 - 36	235	400	235	210
			> 36 - 60	225	400	225	200
			> 60 - 100	205	390	205	185
			> 100 - 150	185	380	185	165
	Q345R (GB/T 713)		3 - 16	345	510	345	310
			> 16 - 36	325	500	325	290
			> 36 - 60	315	490	315	280
			> 60 - 100	305	490	305	275
			> 100 - 150	285	480	285	255

Table 4.0.4 (continued)

Steel type	Grade (Applicable standard)	Delivery state	Steel plate thickness (mm)	Strength at normal temperature		Tensile, compressive, bending strength	
				Yield strength R_e (N/mm^2)	Tensile strength R_m (N/mm^2)	Characteristic value f_{sk} (N/mm^2)	Design value f_s (N/mm^2)
Steel plate for boiler and pressure vessel	Q370R (GB/T 713)	Normalized	10 - 16	370	530	370	330
			> 16 - 36	360	530	360	325
			> 36 - 60	340	520	340	305
Low welding crack susceptibility high-strength steel plate	Q460CF (YB/T 4137)	TM-CP/TM-CP+ tempered or quenched plus tempered	≤ 50	460	550	385	345
			> 50 - 100	440			
	Q500CF (YB/T 4137)		≤ 50	500	610	425	380
			> 50 - 100	480			
	Q550CF (YB/T 4137)		≤ 50	550	670	470	420
			50 - 100	530			
	Q620CF (YB/T 4137)		≤ 50	620	710	495	445
			50 - 100	600			
	Q690CF (YB/T 4137)		≤ 50	690	770	540	485
			50 - 100	670			

Table 4.0.4 (continued)

Steel type	Grade (Applicable standard)	Delivery state	Steel plate thickness (mm)	Strength at normal temperature		Tensile, compressive, bending strength	
				Yield strength R_e (N/mm^2)	Tensile strength R_m (N/mm^2)	Characteristic value f_{sk} (N/mm^2)	Design value f_s (N/mm^2)
Steel for penstocks of hydropower stations	Q345S (GB/T 31946)	Hot rolled, control rolled or normalized	12 - 50	345	490	345	310
			50 - 100	305	490	305	275
			100 - 150	285	480	285	255
	Q490S (GB/T 31946)	Quenched + tempered or TMCP + tempered	12 - 50	490	610	425	380
			50 - 100	470	590	415	375
			100 - 150	450	570	400	360
	Q560S (GB/T 31946)		12 - 50	560	690	485	435
			50 - 100	540	670	470	420
	Q690S (GB/T 31946)		12 - 50	690	780	545	490
			50 - 100	670	760	530	480

NOTE The design value of strength f_s equals the characteristic value of strength f_{sk} divided by the material property partial factor γ_m, and the value of γ_m shall be taken in accordance with the current sector standard NB/T 35056, *Design Code for Steel Penstocks of Hydroelectric Stations*.

5 Basic Design Requirements

5.0.1 The design of steel bifurcation structures for ultimate limit states shall be in accordance with the current national standard GB 50199, *Unified Standard for Reliability Design of Hydraulic Engineering Structures*. The structural safety class of steel bifurcations and corresponding structure importance factor γ_0 shall be in accordance with Table 5.0.1. The structural safety class of steel bifurcations may be adjusted by one level according to the importance or *HD* value of the steel bifurcation, but shall not be inferior to Class II.

Table 5.0.1 Structural safety class and related structure importance factor γ_0 for steel bifurcations

Grade of hydraulic structure	Structural safety class of steel bifurcation	Structure importance factor γ_0
1	I	1.1
2, 3	II	1.0

5.0.2 The design situations of steel bifurcation structure can be classified as persistent, transient or accidental, each of which shall be subjected to ultimate limit states design. The value of the design situation factor ψ shall be taken as follows:

1 1.0 for persistent design situation.

2 0.9 for transient design situation.

3 0.8 for accidental design situation.

5.0.3 The structure factor of steel bifurcations γ_d shall be determined as per Table 5.0.3 according to stress area, location, and type.

Table 5.0.3 Structure factor of steel bifurcations γ_d

Stress type	Stress area and position	Structure factor γ_d
General membrane stress	Pipe wall in membrane stress area	1.50
Local membrane stress	Rib plate	1.35
Local membrane stress	Middle plane of pipe shell within $3.5\sqrt{rt}$ from rib plate and pipe wall at turning point	1.20

Table 5.0.3 *(continued)*

Stress type	Stress area and position	Structure factor γ_d
Local membrane stress + bending stress	Surface of pipe shell within $3.5\sqrt{rt}$ from rib plate and pipe wall at turning point	1.10

NOTES:

1. The γ_d in the table is applicable to the case of weld factor $\varphi = 0.95$; if $\varphi = 0.95$, the γ_d shall be multiplied by $0.95/\varphi$.
2. In the event of hydrostatic test, the γ_d value shall be decreased by 10 % in accordance with the provisions regarding structure factors of exposed steel bifurcations in NB/T 35056, *Design Code for Steel Penstocks of Hydroelectric Stations*.
3. The r and t represent the common-tangent sphere radius and wall thickness of the corresponding cone pipes, respectively.

5.0.4 The weld non-destructive testing rate and weld coefficient φ shall be determined in accordance with Table 5.0.4.

5.0.5 The actions for structural design of steel bifurcations and their partial factors shall be determined in accordance with Table 5.0.5.

Table 5.0.4 Weld non-destructive testing rate and weld factor φ

Non-destructive testing method		RT or TOFD		UT	
Weld type		I	II	I	II
Testing rate (%)	Carbon steel and low-alloy steel	25	10	100	50
	High-strength steel	40	20	100	100
Weld factor φ	Double-side butt welding	0.95			

NOTES:

1. According to project-specific conditions, the testing requirements in design shall not be inferior to the standard in Table 5.0.4. The weld non-destructive testing may use RT (radiographic testing) or UT (ultrasonic testing); RT may be replaced by TOFD (time of flight diffraction) but MT (magnetic particle testing) needs to be conducted. When UT is used, the defective waveforms shall be recorded for further reference.
2. Combined welds shall be subjected 100 % to double-sided, bilateral and multi-angle UT.
3. If UT is used for Class I welds, RT or TOFD shall be used for check. The check length shall not be less than 10 % of the total length of welds; when RT is used for check, each weld shall be provided with at least one piece of film; T-adapter welds and the combined welds of rib plates and pipe shells shall be 100 % checked.
4. When suspicious waveforms are found by UT in Class II welds and cannot be accurately judged, RT or TOFD shall be used for check with a length not less than 5 % of the total length of welds.
5. High-strength steels shall be subjected to surface testing. High-strength steels refer to the tempered or non-tempered steels with $R_e \geq 440$ N/mm² and $R_m \geq 550$ N/mm².

Table 5.0.5 Actions for structural design of steel bifurcations and their partial factors

S/N		Action	Partial factor for action
(1)	(1a)	Internal water pressure	Maximum pressure under normal operation condition (Hydrostatic pressure + water hammer pressure under normal operation condition)	Hydrostatic pressure γ_Q=1.0 Water hammer pressure γ_Q=1.1
	(1b)		Highest pressure under special operation condition (Hydrostatic pressure + water hammer pressure under special operation condition)	Hydrostatic pressure γ_Q=1.0 Water hammer pressure γ_Q=1.1
	(1c)		Internal water pressure for hydrostatic test	γ_Q=1.0
(2)		Full water weight in pipe	γ_Q=1.0
(3)		Structure dead weight	γ_Q=1.05
(4)		Grouting pressure	γ_Q=1.3
(5)		External water pressure	γ_Q=1.0
(6)		Air pressure difference caused by air vent blocking when emptying the pipe	γ_Q=1.0

NOTES:
1. The operation conditions of (1a) and (1b) respectively correspond to the design condition and check condition for water hammer pressure calculation in the current sector standard NB/T 35056, *Design Code for Steel Penstocks of Hydroelectric Stations*.
2. The γ_G and γ_Q in the table are the partial factors for permanent actions and variable actions, respectively.
3. The air pressure difference caused by air vent blocking when emptying the pipe in (6) shall not be smaller than 0.05 N/mm^2 and not greater than 0.1 N/mm^2.

5.0.6 In the verification for hoisting of steel bifurcations members, the dead weight shall be considered in the dynamic factor, which may be taken as 1.2 and may also be adjusted according to the actual situations.

5.0.7 The design situations and action effect combinations of steel bifurcations shall be determined in accordance with Table 5.0.7.

Table 5.0.7 Design situations and action effect combinations of steel bifurcations

Design situation	Action effect combination		Calculation condition
	Combination type	Combinations	
Persistent	Fundamental	(1a)	Normal operation condition
Transient	Fundamental	(1c) + (2) + (3)	Hydrostatic test
		(3) × dynamic factor	Hoisting
		(4)	Construction
		(5) + (6)	Emptying
Accidental	Accidental	(1b)	Special operation condition

NOTE The action serial number in combination item shall be in accordance with Table 5.0.5 of this code.

5.0.8 For ultimate limit states, the stress of each calculation point shall meet the following requirements:

$$\sigma \leq \sigma_R \quad (5.0.8\text{-}1)$$

The general expression of σ for fundamental combination is:

$$\sigma = S(\gamma_G G_k, \gamma_Q Q_k, a_k) \quad (5.0.8\text{-}2)$$

The general expression of σ for accidential combination is:

$$\sigma = S(\gamma_G G_k, \gamma_Q Q_k, \gamma_A A_k, a_k) \quad (5.0.8\text{-}3)$$

The stress of each calculation point shall be calculated by the fourth strength theory (von Mises yield criterion), the formula is:

$$\sigma = S(\bullet) = \sqrt{\sigma_\theta^2 + \sigma_x^2 + \sigma_r^2 - \sigma_\theta \sigma_x - \sigma_\theta \sigma_r - \sigma_x \sigma_r + 3\left(\tau_{\theta x}^2 + \tau_{\theta r}^2 + \tau_{xr}^2\right)}$$

$$(5.0.8\text{-}4)$$

Which can be simplified as follows in the case of a plane problem:

$$\sigma = S(\bullet) = \sqrt{\sigma_\theta^2 + \sigma_x^2 - \sigma_\theta \sigma_x + 3\tau_{\theta x}^2} \quad (5.0.8\text{-}5)$$

The σ_R shall be calculated by the following formula:

$$\sigma_R = \frac{1}{\gamma_0 \psi \gamma_d} f \quad (5.0.8\text{-}6)$$

where

σ	is the design value of action combination of structural members of steel bifurcations expressed by penstock stress (N/mm^2);
σ_R	is the resistance limit of structural members of steel bifurcation (N/mm^2);
$S(\cdot)$	is the action combination function for structural members of steel bifurcation;
γ_G, γ_Q	are the partial factors for permanent action and variable action, determined as per Table 5.0.5;
γ_A	is the partial factor for accidental action, and shall be taken as 1.0;
G_k, Q_k	are the characteristic values of permanent action and variable action;
A_k	is the representative value of accidental action;
a_k	is the geometric parameter characteristic value of structural members of penstock;
γ_0	is the structural importance factor, determined as per Table 5.0.1;
ψ	is the design situation factor, determined as per Table 5.0.2;
γ_d	is the structure factor, determined as per Table 5.0.3;
f	is the design value of steel strength (N/mm^2), determined as per Table 4.0.4;
σ_x	is the axial normal stress (N/mm^2), positive for tension;
σ_θ	is the circumferential normal stress (N/mm^2), positive for tension;
σ_r	is the radial normal stress (N/mm^2), positive for tension;
$\tau_{\theta x}, \tau_{\theta r}, \tau_{xr}$	are the shear stresses (N/mm^2).

5.0.9 When subjected to uniformly distributed external pressure, the stability against external pressure can be verified by the following formula:

$$p_{0k} \leq \frac{p_{cr}}{K_c} \tag{5.0.9}$$

where

p_{0k} is the characteristic value of radial uniform external pressure (N/mm^2);

p_{cr} is the critical external buckling pressure (N/mm^2);

K_c is the buckling safety factor against external pressure, which is taken as 2.0 for smooth pipe wall, and 1.8 for the pipe wall between stiffener rings and for the stiffener ring.

6 Selection of Shape Parameters

6.0.1 The shape design of steel bifurcations shall consider the structural and hydraulic characteristics.

6.0.2 The diameter of branch pipe should be determined such that the flow velocity in the branch pipe is equal to that in the main pipe.

6.0.3 The bifurcating angle may be selected in a range of 55° to 90°. Bifurcating angle should be 55° to 70° for asymmetric Y-shaped steel bifurcations and 65° to 80° for symmetric Y-shaped steel bifurcations.

6.0.4 The amplification ratio of steel bifurcations should range from 1.1 to 1.2.

6.0.5 Single or multiple cones may be used before and after the bifurcating point. When the HD value of steel bifurcations is greater than 1,500 m · m, multiple cones should be adopted. For the multi-cone steel bifurcations (Figure 6.0.5), the turning angles of main and branch cone waistlines should be gentle, and bifurcating angles shall be reasonably distributed so as to make the stress distribution at the corner points relatively uniform. The limit values of turning angles of main and branch cone waistlines are shown in Table 6.0.5. The shape of steel bifurcations should be determined by comparison of alternatives and the steel bifurcation with smaller size should be selected. For multi-cone steel bifurcations, C_0 should take the smaller value and C_{22} should take the larger value.

Table 6.0.5 Limit values of turning angles of main and branch cone waistlines

Description	Single-cone steel bifurcation			Double-cone steel bifurcation				
	C_1	C_0	C_2	C_{11}	C_1	C_0	C_2	C_{22}
Symmetric Y-shape	14°	13°	20°	8°	8°	8°	11°	11°
Asymmetric Y-shape	19°	19°	20°	13°	13°	12°	15°	18°

NOTE For a single-cone steel bifurcation, if the wall thickness of the steel bifurcation is the same, C_2 may be larger but should not exceed 25°.

6.0.6 The rib-width ratio of rib plates may be selected in the range of 0.20 to 0.50. The rib-width ratio should be 0.30 to 0.45 for asymmetric Y-shaped steel bifurcations and 0.25 to 0.35 for symmetric Y-shaped steel bifurcations.

6.0.7 The inner edge curve of rib plates should be preferably elliptic curve and its equation be optimized by finite element calculation; parabola may also be used.

6.0.8 After preliminary selection of the shape parameters, the shape of the steel bifurcation may be determined by the methods given in Appendix A of this code and the important steel bifurcations should be subjected to finite element structural calculation for optimization.

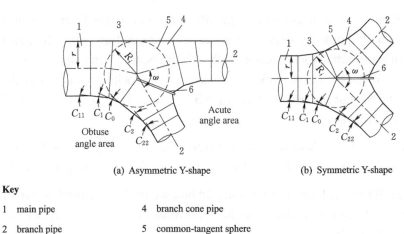

(a) Asymmetric Y-shape (b) Symmetric Y-shape

Key

1 main pipe
2 branch pipe
3 main cone pipe
4 branch cone pipe
5 common-tangent sphere
6 crescent rib

Figure 6.0.5 Shape of multi-cone steel bifurcation

6.0.9 The head loss coefficient of symmetric Y-shaped steel bifurcations may be determined preliminarily in accordance with Table 6.0.9, that of asymmetric ones may be determined preliminarily in accordance with the current sector standard NB/T 35021, *Design Code for Surge Chamber of Hydropower Stations*, and the final head loss should be determined by hydraulic numerical calculation or model test.

Table 6.0.9 Head loss coefficient ξ for symmetric Y-shaped steel bifurcations

Flow pattern	Bifurcating angle	55°	75°	90°
Double-side flow	Dividing flow	0.12	0.21	0.25
	United flow	0.24	0.25	0.36
Single-side flow	Dividing flow	0.50 - 1.60		
	United flow	0.50 - 1.40		

NOTES:
1 The head loss coefficient in the table corresponds to the flow velocity in the main pipe.
2 The head loss coefficient for the steel bifurcation with other bifurcating angle may be calculated by interpolation based on the value given in the table.

7 Structural Design

7.0.1 The structural analysis method of underground crescent-rib reinforced bifurcation should comply with Appendix B of this code.

7.0.2 When the design considers the situation where the internal water pressure is jointly borne by steel bifurcations and surrounding rocks, the cover thickness of surrounding rocks shall meet the requirements given in Article B.1.2 of this code. When the cover thickness of surrounding rocks does not meet the requirements, the current sector standard NB/T 35056, *Design Code for Steel Penstocks of Hydroelectric Stations* shall apply.

7.0.3 The structural design of steel bifurcations bearing internal water pressure jointly with surrounding rocks shall meet the following requirements:

1. The 3D finite element method should be used in the structural design of steel bifurcations to simulate the constraints of surrounding rocks and gap reasonably.

2. The design of steel bifurcations sharing internal water pressure with surrounding rocks shall satisfy the exposed bifurcation criterion.

3. The surrounding rock pressure in cavern section of steel bifurcations shall be borne jointly by cavern support and backfill concrete instead of steel bifurcations.

7.0.4 The engineering geological conditions of the surrounding rocks of cavern section of steel bifurcations shall be ascertained, the elastic modulus, deformation modulus, unit elastic resistance coefficient, Poisson's ratio and other geomechanical parameters of the surrounding rocks should be determined by field tests, and the values used in calculation shall consider the influence of blasting loosening, adjacent caverns and free surfaces.

7.0.5 In the structural calculation considering the joint action of steel bifurcation and surrounding rocks, the gap may be calculated taking the steel bifurcation as an underground cylindrical pipe with the same diameter as the common-tangent sphere, as per Appendix B of this code. The ratio of the smallest gap of steel bifurcation to the radius of the common-tangent sphere should not be less than 4×10^{-4}, and the vertical gap and horizontal gap may be different.

7.0.6 The groundwater pressure acting on the steel bifurcation shall be determined through comprehensive analysis of investigation data, influences of the reservoir seepage and water conveyance system leakage and effects of drainage system.

7.0.7 The thermal actions on steel bifurcations may be ignored.

7.0.8 The structural design of steel bifurcations should take the following steps:

1 As per Appendix B of this code, calculate the average sharing ratio of surrounding rocks for the steel bifurcation as an underground cylindrical pipe whose diameter equals the diameter of the common-tangent sphere of the steel bifurcation; determine the internal water pressure shared by the steel bifurcation, based on which calculate the pipe wall thickness by membrane stress and local stress, respectively, and take the larger value. The pipe wall thickness can be calculated by the following formulae:

By membrane stress:

$$t_{y1} = \frac{k_1 p_1 r_i}{\sigma_{R_1} \cos A} \tag{7.0.8-1}$$

By local stress:

$$t_{y2} = \frac{k_2 p_1 r_i}{\sigma_{R_2} \cos A} \tag{7.0.8-2}$$

where

t_{y1}, t_{y2} are the wall thicknesses of steel bifurcations estimated by membrane stress and local stress, respectively (mm);

p_1 is the internal water pressure on steel bifurcations (N/mm²);

r_i is the radius of rotation from this pipe shell calculation point to the axis of rotation (i.e. vertical distance), which is the steel pipe radius for isodiametric pipes (mm);

A is the half-apex angle of this steel pipe section (°);

k_1 is the stress concentration coefficient at the corner point of waistline, taken as 1.0 to 1.1 for crescent-rib reinforced bifurcation;

k_2 is the stress concentration coefficient at the corner point of waistline, selected from Figure 7.0.8;

$\sigma_{R_1}, \sigma_{R_2}$ are the resistance limits calculated by the membrane stress and by the local membrane stress plus bending stress, respectively (N/mm²).

2 Rib plate thickness may be estimated as 1.9 to 2.3 times the maximum wall thickness of the steel bifurcation.

3　The finite element analysis and structural optimization design should be carried out.

7.0.9　The finite element structural analysis of steel bifurcations shall meet the following requirements:

1　Rib plates should be simulated by solid elements; pipe shells should be simulated by shell elements; in the case of buried pipes, the joint actions of steel bifurcation and surrounding rocks shall be simulated by contact elements.

2　The ratio of the maximum size of finite elements to the radius of common-tangent sphere should not be greater than 0.10.

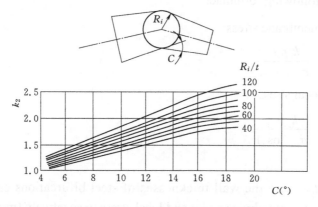

Figure 7.0.8　Stress concentration coefficient k_2 at the corner point of waistline

3　For the finite element model under operating condition, the ratio of the length of straight pipe section of main or branch pipe to the radius of main or branch pipe should be 3 to 4.

4　The pipe-end boundary for finite element model under operating condition should be fully constrained or be displacement constrained in radial, circumferential and axial directions.

5　Bulkheads and supports should be simulated under the hydrostatic test condition, taking into account the water weight and structure self-weight.

7.0.10　The critical external buckling pressure of the steel bifurcation may be calculated as per Appendix B.2 of this code.

8 Detailing Requirements

8.0.1 The minimum wall thickness of a steel bifurcation with an allowance shall not only satisfy the bearing capacity but also the stiffness required by manufacture, transportation and installation, which should be calculated by the following formula but should not be less than 6 mm.

$$t \geq D/800 + 4 \tag{8.0.1}$$

where

t is the wall thickness of the steel bifurcation, which shall be rounded up to an integer (mm);

D is the diameter of the common-tangent sphere of the steel bifurcation (mm).

8.0.2 A large steel bifurcation may be designed as varying wall thickness and the wall thickness difference between adjacent cones of a bifurcation should not be greater than 4 mm; if greater than 4 mm, the joint face of the thicker steel plate shall be beveled to 1:3.

8.0.3 The spacing of circumferential welds of steel bifurcation should not be less than 10 times the pipe wall thickness and should not be less than 300 mm.

8.0.4 Longitudinal welds shall not be arranged on the horizontal or vertical axis of the cross section of the steel bifurcation, and shall have an angle greater than 10° to the horizontal or vertical axis and an arc distance greater than 300 mm and 10 times the pipe wall thickness from the horizontal or vertical axis. The butt weld of rib plates shall not be arranged on the symmetric axis of the rib, and the included angle between the weld and the symmetric axis shall be greater than 15°. The spacing between the longitudinal joints of adjacent pipe segments shall also meet the above requirements.

8.0.5 The following welds of steel bifurcations shall be of Class I and other load-carrying welds shall be of Class II.

1. Longitudinal and circumferential welds on pipe walls.

2. Butt welds on reinforced rib plates.

3. Composite welds on butt and fillet welding at the junction of reinforced rib plates and pipe shells.

4. Bulkhead welds and the welds connecting the bulkhead and the pipe wall.

5. Butt welds on stiffener rings.

8.0.6 The fabrication deviation of steel bifurcations shall comply with the current national standard of GB 50766, *Code for Manufacture Installation and Acceptance of Steel Penstocks in Hydroelectric and Hydraulic Engineering*, and the pipe roundness deviation after installation shall not be greater than 5D/1000 but not exceed 40 mm. At least two pairs of pipe diameters shall be measured for the roundness deviation. If the stiffness of a steel bifurcation does not meet the requirements for hoisting and encasement concrete pouring, necessary reinforcement measures shall be taken inside and outside the steel bifurcation.

8.0.7 The pre-heating of steel bifurcations before welding shall comply with the current national standard of GB 50766, *Code for Manufacture Installation and Acceptance of Steel Penstocks in Hydroelectric and Hydraulic Engineering*.

8.0.8 In any of the following cases, the steel bifurcation shall be stress relieved:

1. The shell thickness exceeds the following values: 42 mm for Q235 and Q245R, 38 mm for Q345, and 36 mm for Q390.

2. The thickness (t) of steel plates for cold formed pipe segments exceeds the following values: D/33 for Q235 and Q345 and D/40 for Q390.

3. For case 1, when the steel bifurcation is complex in shape and is difficult to treat as a whole, stress relief may only cover the welds.

4. For other steel types requiring stress relief, a special study shall be conducted. For high-strength steels, the residual stress should be reduced by using reasonable welding process and strict quality control measures, and should not be eliminated by heat treatment. If the residual stress is to be eliminated by explosion or vibration aging, process test shall be conducted before fabrication of steel bifurcation to determine reasonable parameters, so as to ensure the mechanical properties of welds and the effect of residual stress relief.

8.0.9 Steel bifurcations should be fabricated integral for installation. The split steel bifurcation should have as few segments as possible to minimize field welding and shall be preassembled in shop before delivery.

8.0.10 Considering corrosion and wear, the wall thickness allowance of the steel bifurcation may be taken as 2.0 mm. For the steel bifurcation subjected to severe corrosion and sediment erosion, the thickness allowance shall be demonstrated.

8.0.11 The corrosion control design of steel bifurcations shall be determined considering the flow velocity and quality of water, content and properties of sediment, environment, groundwater quality, etc. The corrosion control scheme, surface pre-treatment, construction processes and quality inspection shall comply with the current national standard GB 50766, *Code for Manufacture Installation and Acceptance of Steel Penstocks in Hydroelectric and Hydraulic Engineering*.

8.0.12 The backfill concrete between the underground steel bifurcation and the surrounding rocks shall be compacted, and the locations of the limbs, stiffener rings, and crotch of the steel bifurcation must be well vibrated to minimize the gap between the bifurcation and the concrete and between the concrete and the surrounding rocks. The strength grade of concrete backfilled outside steel bifurcations shall not be inferior to C20. Micro-expansive concrete or concrete with low drying shrinkage may be used after study. For the positions difficult to vibrate, self-compacting concrete may be used.

8.0.13 The top of the underground steel bifurcation shall be backfill grouted. The grouting pressure shall not be less than 0.2 MPa nor greater than the design external pressure resistance of the steel bifurcation.

8.0.14 The bottom of the steel bifurcation and both sides of rib plates shall be contact grouted. The grouting pressure should not be greater than 0.2 MPa and shall ensure that the bifurcation deformation during contact grouting does not exceed the allowable design value. Contact grouting should be conducted at lower temperature.

8.0.15 When a bifurcation is made of high-strength steels, embedded grouting piping system should be used for backfill grouting and contact grouting, and reliable quality assurance measures shall be taken.

8.0.16 Consolidation grouting shall be conducted after study based on the geological conditions of surrounding rocks, internal water pressure, design assumptions, and excavation blasting methods. Consolidation grouting should be conducted before installation of bifurcations or using the embedded grouting piping system. Grouting holes should not be drilled on steel bifurcations. The consolidation grouting pressure should not be less than 0.5 MPa. If embedded grouting pipe is adopted, safety monitoring shall be conducted during grouting to prevent buckling of bifurcations.

8.0.17 Asymmetric Y-shaped steel bifurcations may be provided with guide plates at rib plates of interior walls, and the layout and pattern of the guide plates should be determined by hydraulic model test. The guide plates should be set with pressure-balancing holes and the constraints on pipe shells shall be decreased.

8.0.18 The distance between the outer edge of rib plate and the outer wall of pipe shell shall meet the detailing requirements for welding, and should not be less than 50 mm at the rib waist.

9 Testing

9.1 Model Test

9.1.1 Steel bifurcations may be subjected to monolithic structure model test. When new materials, new design methodology, or new processes are used for the steel bifurcations, structural model test should be performed.

9.1.2 The scale, range and action of the model test shall be determined according to the test purpose.

9.1.3 The structural model test of steel bifurcation should use the structural model made of simulation materials.

9.1.4 For the structural model test of steel bifurcation, the internal water pressure of the prototype at overall yield, which is converted from the measured data of the model, shall not be less than 2.0 times the design internal water pressure of the prototype. If the pipe bursting test is required, the safety factor for measured internal water pressure of pipe bursting shall not be less than 3.0.

9.1.5 When the structural model test results are cited, the safety of the structure shall be defined based on the difference between the model and the actual structure in scale, materials, constraint conditions and similarity conditions.

9.1.6 Hydraulic model test should be conducted for Grade 1 or 2 steel bifurcations with complex shapes to verify the rationality of the shapes of steel bifurcation and to make adjustment according to the test result.

9.1.7 In the hydraulic model test, a larger model scale should be selected to reduce the influence of flow pattern dissimilarity on hydraulic characteristics. Under normal operation condition, the Reynolds number for the bifurcation model should not be less than 3×10^5.

9.2 Hydrostatic Test

9.2.1 The steel bifurcation shall undergo hydrostatic test in the absence of demonstration and strict process and quality control.

9.2.2 Hydrostatic test for the steel bifurcation should be conducted in shop.

9.2.3 The monitoring items and arrangement for hydrostatic test shall meet the following requirements:

1. The monitoring items may include internal water pressure, water temperature, deformation, stress-strain of pipe shells and rib plates, inflow test under different pressures, etc.

2 For stress-strain monitoring, one monitoring section should be arranged each at the rib plate, main cone, intersecting line between the main and branch cone, and branch cone; and monitoring points shall be set at upper 1/4 or 1/2 of the section generally; in addition, monitoring points shall be set at the waistline corner point, top of steel bifurcations, peak stress zones, rib-side pipe walls, and general membrane stress zones of pipe walls. Sensors should be set at monitoring points on both internal and external surfaces of rib plates and pipe shell.

3 For deformation monitoring, monitoring points shall be set at the top and bottom of steel bifurcations and at the waistline corner points on intersecting line of main cone and branch cone, and should also be set at waistlines of rib plates, bulkheads, etc.

4 The layout of stress-strain monitoring points should correspond to that of prototype monitoring points.

5 Displacement and strain warning values shall be set according to test pressure, and real-time monitoring shall be carried out.

9.2.4 The test pressure shall be selected according to the following requirements:

1 For the steel bifurcations with internal water pressure not shared by surrounding rocks, the test pressure shall be 1.25 times the highest design value of internal water pressure under normal operation condition and shall not be smaller than the highest design value of internal water pressure under special operation condition.

2 For the steel bifurcations with internal water pressure shared by surrounding rocks, the test pressure shall be determined by finite element calculation according to shapes of steel bifurcations, test conditions, and resistance limit of the steel bifurcation under hydrostatic test condition.

3 Water pressure shall be slowly step increased or decreased. The pressure hold time for each step and for the maximum test pressure shall not be less than 30 min. The pressurization or depressurization rate should not be greater than 0.05 MPa/min.

9.2.5 The steel bifurcation should undergo hydrostatic test for two complete cycles.

9.2.6 The steel bifurcations should be subjected to non-destructive testing after the hydrostatic test, and to residual stress test when necessary.

9.2.7 The ambient temperature and test water temperature during hydrostatic test of steel bifurcations shall be above 5 °C.

9.2.8 The supporting structure for hydrostatic test of steel bifurcations shall have sufficient strength, rigidity and stability. The constraints of supporting structure on steel bifurcations should be reduced. The structural safety shall be evaluated considering the difference between the test state and the operation state.

9.2.9 For the steel bifurcation subjected to hydrostatic test, certain cutting margin shall be reserved during fabrication for the main and branch cones connected to test bulkheads.

9.2.10 For hydrostatic test of steel bifurcations, the air at the roof of steel bifurcations shall be exhausted and the exhaust pipes shall be led to the roof of steel bifurcations from test bulkheads of steel bifurcations. Manholes shall be provided in test bulkheads of steel bifurcations.

9.2.11 Safety measures shall be taken during hydrostatic test of steel bifurcations.

10 Safety Monitoring

10.0.1 Steel bifurcations shall be subjected to safety monitoring, which shall be incorporated into the project safety monitoring system.

10.0.2 Monitoring items shall be arranged according to the scale, structural characteristics, and geological conditions of steel bifurcations, taking into account the water heads and operation modes of the power station. The monitoring should include the following items:

1. Deformation and compressive stress of surrounding rocks.
2. Gaps between the steel bifurcation and the backfill concrete, and between the backfill concrete and the surrounding rocks.
3. Stress-strain of pipe shells and rib plates.
4. Internal and external water pressures.
5. Strain of backfill concrete.
6. Temperatures of pipe walls and surrounding rocks.

10.0.3 The monitoring of steel bifurcations should be arranged according to the following requirements:

1. For the deformation of surrounding rocks around the steel bifurcation, 1 or 2 monitoring sections should be set, and multipoint displacement meters should be set, perpendicular to the design excavation line, in the surrounding rocks at the crown and waistline of the steel bifurcation; for the compressive stress in the surrounding rocks, compressive stress meters shall be set on the surface of surrounding rocks at the bottom and waistline of the steel bifurcation.

2. Gaps between the steel bifurcation and the backfill concrete and between the backfill concrete and the surrounding rocks should be monitored, and monitoring sections should be set at the intersecting lines between the transitional cone and the main cone and between the main cone and the branch cone. One set of monitoring points shall be arranged at the waistline for each section, and at least one set of monitoring points shall be arranged at the top or bottom of the steel bifurcation. For each section, monitoring points shall be arranged between the top backfill concrete and the surrounding rocks and between the steel bifurcation bottom and the backfill concrete.

3. Stress-strain shall be monitored by steel plate meters. At least 3 monitoring sections shall be set at the intersecting lines between the

main and branch cones, between the transitional cone and the main cone or pipe, and between the transitional cone and the branch cone or pipe. Moreover, a monitoring section should be set at the centerline of rib plate along its thickness. The layout of monitoring sections and monitoring points is shown in Figure 10.0.3.

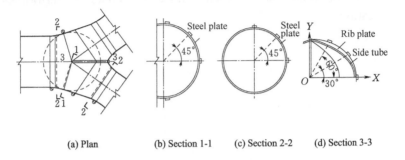

(a) Plan (b) Section 1-1 (c) Section 2-2 (d) Section 3-3

Key

▭ steel plate meter (circumferential, parallel to the flow direction)

○ steel plate meter (circumferential, perpendicular to the flow direction)

Figure 10.0.3 Layout of steel plate meters for steel bifurcations

4   The internal water pressure of steel bifurcations should be monitored using the pressure meters and pressure sensors in front of inlet valves. The external water pressure of steel bifurcations should be monitored by piezometers which shall be laid out on the surrounding rocks and external walls of pipe shells, taking into account the monitoring sections for surrounding rock deformation and pipe shell stress-strain. The piezometers inside surrounding rocks around the steel bifurcation shall have at least 3 monitoring points and those beside external walls of pipe shells should have at least 2 monitoring points. If conditions allow, piezometer tubes may be provided at auxiliary caverns or along water conveyance lines to monitor the external water pressure of steel bifurcations.

5   For monitoring the strain of backfill concrete, two-directional strain meters, radial and circumferential, may be provided in the middle of backfill concretes at the waistline and the top or bottom of the vertical centerline, taking into account the monitoring sections for surrounding rock deformation and pipe shell stress-strain.

6   When the temperature of pipe walls and surrounding rocks is monitored by electrical thermometers, 1 or 2 monitoring points should be set on the external wall of the steel bifurcation, and 1 or 2 monitoring points may be set inside surrounding rocks.

10.0.4 Observation shall commence upon completion of the monitoring instrument installation. The frequency of observation shall be increased during filling and emptying of the water conveyance system and the internal water pressure shall be observed at the same time. Monitoring data, collected during the filling of water conveyance system and the commissioning and operation of units, shall be sorted out and analyzed in time to evaluate the performance of steel bifurcations.

Appendix A Shape Design Methods for Underground Crescent-Rib Reinforced Bifurcations

A.0.1 The shape calculation of underground crescent-rib reinforced bifurcations shall be conducted by the following steps:

1. Initially define the turning angles of waistlines C, D, W_L and the largest common-tangent sphere diameter, and draft the shape sketch. The proposed shape shall be coordinated with the plant layout. The three ends shall be cylindrical pipes and the two waistlines in parallel.

2. Calculate angles A and B of each segment, preferably starting with the three ends. Number the main pipe segment i along the water flow while number the two branch pipe segments against the water flow.

3. Calculate the waistline length l_i and the axis length m_i of each segment, and the radius R_i of common-tangent sphere. The radius may be obtained by determining the length of waistlines first or the length of waistlines may be obtained by determining the radius first. The two waistlines must match the radius, and there shall be one common-tangent sphere where three cones intersect.

4. Steel bifurcation plan shall be prepared and its rationality shall be checked.

5. Calculate the coordinates E and F of the intersecting line between the middle plane of pipe wall and the middle plane of rib plate, the coordinate of the intersecting point Q of the three cone pipes, and the miter cut angles K_2 and K_3. Calculate the rib-width b_r according to the given pipe wall thickness t and rib plate thickness t_r, plot the rib plate outline and check its rationality.

6. The calculations shall include the following:

    1) Obtain the coordinates E_0 and F_0 of intersecting line between pipe shell external surface and rib plate side face for the assembly of steel bifurcations.

    2) Calculate the bottom circle radii R_0 and R_2, the pointed end coordinates x, y and y_2, the nodal increment C_3 and D_3, and the bottom line coordinates x_1 and y_1, for formulation of segment diagram and developed view.

    3) Check the pipe wall thickness.

4) Calculate the side-line length L_1 of expanded sector, the central angle S, the difference between large end L_1 and small end L_1, and the waistline length D_L occupied by rib thickness. Check the consistency between the side-line length of developed view and the waistline length.

5) Calculate the areas F_4 and F_5 of pipe segments for the calculation of work quantities.

A.0.2 Geometric dimensions shall be calculated according to the following requirements:

1 The geometric dimension of steel bifurcations shall be calculated according to the internal surface of pipe walls, except otherwise indicated.

1) The half-apex angle and axis turning angle (Figure A.0.2-1) can be calculated by the following formulae:

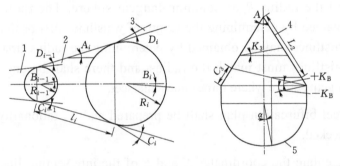

(a) Cone pipe dimensions   (b) Developed view of cone pipe

Key
1 segment $i-1$ 4 cone pipe
2 segment i 5 bottom circle
3 segment $i+1$

Figure A.0.2-1   Half-apex angle and axis turning angle

$$A_{i+1} = A_i + \frac{C_i + D_i}{2} \quad \text{(A.0.2-1)}$$

$$B_i = \frac{C_i - D_i}{2} \quad \text{(A.0.2-2)}$$

where

A_{i+1} is the half-apex angle of segment $i+1$ (°);

A_i is the half-apex angle (°) of segment i; the three-end half-apex angle is known to be zero and used as the starting point of calculation;

C_i is the turning angle of the right waistline of segment i (°);

D_i is the turning angle of the left waistline of segment i (°);

B_i is the axis turning angle of segment i (°).

2) The calculations of waistline length and axis length shall meet the following requirements.

Waistline length of general pipe segments can be calculated by the following formula:

$$l_i = R_i\left(\cot A_i - \tan\frac{C_i}{2}\right) - R_{i-1}\left(\tan\frac{C_{i-1}}{2} + \cot A_i\right) \qquad (A.0.2\text{-}3)$$

where

l_i is the waistline length of segment i (mm);

R_i is the radius of common-tangent sphere of segment i (mm).

The axis length can be calculated by the following formula:

$$m_i = (R_i - R_{i-1}) / \sin A_i \qquad (A.0.2\text{-}4)$$

where

m_i is the axis length of segment i, i.e. the distance between centers of two common-tangent spheres (mm).

The waistline length of three-cone intersection (Figure A.0.2-2) can be calculated by the following formula:

$$l_i = R_{i-1}\left(\cot A_i - \tan\frac{W_L}{2}\right) - R_i\left(\cot A_i + \tan\frac{C_i}{2}\right) \qquad (A.0.2\text{-}5)$$

where

W_L is the intersecting angle of waistlines of two branch cone pipes (°).

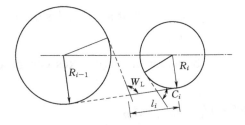

Figure A.0.2-2　Intersecting of three cones

The axis length of the first pipe segment (cylindrical pipe) connecting to the three ends of steel bifurcation (Figure A.0.2-3) can be calculated by the following formula:

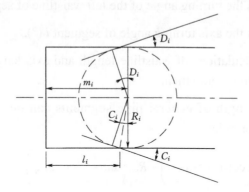

Figure A.0.2-3 Axis length of the first pipe segment

$$m_i = l_i + R_i \tan \frac{C_i}{2} \quad (A.0.2\text{-}6)$$

3) The miter cut angles at intersecting point of three cones can be calculated by the following formulae:

$$\tan K_2 = \frac{\cos A_{N1} \sin \omega_2}{\cos A_{N2+1} - \cos A_{N1} \cos \omega_2} \quad (A.0.2\text{-}7)$$

$$\tan K_3 = \frac{\cos A_{N1} \sin \omega_3}{\cos A_{N1+1} - \cos A_{N1} \cos \omega_3} \quad (A.0.2\text{-}8)$$

where

K_2 is the miter cut angle at the left side of cone pipe bottom line (°);

K_3 is the miter cut angle at the right side of cone pipe bottom line (°);

ω_2, ω_3 are the intersecting angles between the axes at the bifurcating point (°);

A_N, A_{N1+1}, A_{N2+1} are the half-apex angles of three-cone intersection (°).

The intersecting angles between the three axes can be calculated by the following formulae:

$$\omega_1 = 180° - W_L - A_{N1+1} - A_{N2+1} \quad (A.0.2\text{-}9)$$

$$\omega_2 = 180° - C_{N1} - A_{N1} - A_{N2+1} \quad (A.0.2\text{-}10)$$

$$\omega_3 = 180° - D_{N1} - A_{N1} - A_{N1+1} \quad (A.0.2\text{-}11)$$

where

ω_1 — is the intersecting angle between the axes at the bifurcating point (°);

W_L, C_{N1}, D_{N1} — are the intersecting angles between waistlines of adjacent cone pipes at the three-cone intersection (°).

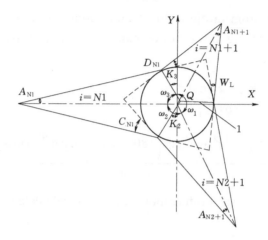

Key

1 rib plate

Figure A.0.2-4   Three-cone intersection plan

4) The coordinates x and y of intersecting point Q of the three-cone intersecting lines can be calculated by the following formulae:

$$\frac{x}{R_{N1}} = \frac{A_{P7} + A_{P8}}{A_{P9} + A_{P10} + A_{P11}} \qquad (A.0.2\text{-}12)$$

$$\frac{y}{R_{N1}} = \frac{A_{P12} + A_{P13} + A_{P14}}{A_{P9} + A_{P10} + A_{P11}} \qquad (A.0.2\text{-}13)$$

$$A_{P7} = \sin(A_{N1+1} - A_{N1})\sin\omega_2 \qquad (A.0.2\text{-}14)$$

$$A_{P8} = \sin(A_{N2+1} - A_{N1})\sin\omega_3 \qquad (A.0.2\text{-}15)$$

$$A_{P9} = \cos A_{N1}\sin\omega_1 \qquad (A.0.2\text{-}16)$$

$$A_{P10} = \cos A_{N1+1}\sin\omega_2 \qquad (A.0.2\text{-}17)$$

$$A_{P11} = \cos A_{N2+1}\sin\omega_3 \qquad (A.0.2\text{-}18)$$

$$A_{P12} = \sin(A_{N1} - A_{N1+1})\cos\omega_2 \qquad (A.0.2\text{-}19)$$

$$A_{P13} = \sin(A_{N2+1} - A_{N1})\cos\omega_3 \qquad (A.0.2\text{-}20)$$

$$A_{P14} = \sin(A_{N1+1} - A_{N2+1}) \qquad (A.0.2\text{-}21)$$

where

- x, y are the coordinates of three-cone intersecting point, taking the center of common-tangent sphere as the origin;
- R_{N1} is the radius of common-tangent sphere of segment $N1$ (mm).

5) The bottom circle radii of main cone and branch cones of the intersecting three cones can be calculated by the following formulae:

$$R_{0N1} = \frac{R_{N1} + x \sin A_{N1}}{\cos A_{N1}} \quad (A.0.2\text{-}22)$$

$$R_0 = \frac{R_{N1} - \sqrt{x^2 + y^2} \cos\left(180° - \omega_3 - \arctan\frac{y}{x}\right) \sin A_{N1+1}}{\cos A_{N1+1}} \quad (A.0.2\text{-}23)$$

For symmetric steel bifurcations, R_0 can be calculated by the following formula:

$$R_0 = \frac{R_{N1} - x \cos\frac{\omega_1}{2} \sin A_{N1+1}}{\cos A_{N1+1}} \quad (A.0.2\text{-}24)$$

where

- R_{N1} is the bottom circle radius of main cone segment $N1$ (mm)
- R_0 is the bottom circle radius of branch cone segment $N1+1$ (mm);
- x, y are the coordinates of three-cone intersecting point, taking the center of common-tangent sphere as the origin.

6) The coordinates E and F of the intersecting line (Figure A.0.2-5) can be calculated by the following formulae:

$$E = \frac{R_0 (\sin \alpha - \sin \alpha_0)}{(1 + A_0 \sin \alpha) \cos K_3} \quad (A.0.2\text{-}25)$$

$$F = \frac{R_0 A_4 \cos \alpha}{1 + A_0 \sin \alpha} \quad (A.0.2\text{-}26)$$

$$\sin \alpha_0 = \frac{y}{R_0} \quad (A.0.2\text{-}27)$$

$$A_0 = \tan A \cot \varphi_1 \quad (A.0.2\text{-}28)$$

$$A_4 = 1 + A_0 \sin \alpha_0 \quad (A.0.2\text{-}29)$$

where

A	is the half-apex angle of cone pipe (°);
φ_1	is the inclination angle of intersecting line of branch cone pipes (°);
α_0	is the central angle of bottom circle corresponding to the rib end point (°);
α	is the central angle of bottom circle (°);
y	is the distance between the rib end point and the center of bottom circle of segment $N1+1$ (mm);
A_0, A_4	are parameters;
K_3	is the included angle between the rib plate centerline and the radius of bottom circle of segment $N1+1$ (°).

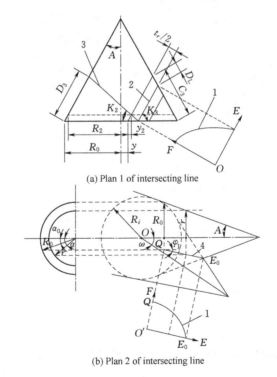

(a) Plan 1 of intersecting line

(b) Plan 2 of intersecting line

Key

1　intersecting line of middle plane of rib wall

2, 4　rib plates

3　bottom line

Figure A.0.2-5　Calculation diagram for coordinates of intersecting line

2　The calculation of geometric dimensions of rib plates shall meet the following requirements:

1) Calculate horizontally projected length E_m of intersecting line according to the coordinates of intersecting line, and consult the rib plate width reference curve (Figure A.0.2-6) for the initial rib-width ratio β according to the bifurcating angle ω. The width b_{r0} of waist section can be calculated by the following formula:

$$b_{r0} = \beta E_m \qquad (A.0.2\text{-}30)$$

where

b_{r0} is the width of waist section, i.e. the horizontal distance from internal edge of the rib plate to the intersecting line;

β is the rib-width ratio;

E_m is the horizontally projected length of the intersecting line.

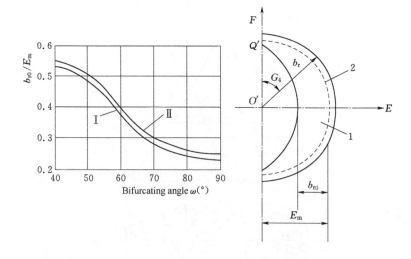

Key

1   rib plate
2   intersecting line of pipe shells
I   test conditions
II   operation conditions

Figure A.0.2-6   Rib plate width reference curves

2) Rib plate thickness may be proposed as 1.9 to 2.3 times the maximum wall thickness of the steel bifurcation.

3) The major and minor axes of the elliptic equation of internal and external surfaces intersecting line of the pipe shell (Figure A.0.2-7) and coordinates of the elliptic center can be obtained by the geometric method, thus obtaining the elliptic equation of

intersecting line with the intersecting point Q as the origin. The elliptic equation is shown as follows:

$$\frac{(x-x_0)^2}{a^2} + \frac{(y-y_0)^2}{b^2} = 1 \qquad (A.0.2\text{-}31)$$

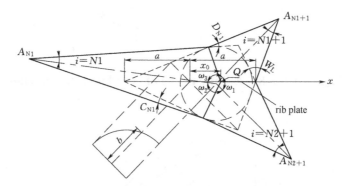

Figure A.0.2-7 Calculation diagram of internal and external surface intersecting lines of pipe shells

4) Determine y_b of the top point A of rib plate internal edge and x_a of point B of rib plate waist internal edge according to rib plate welding requirements, waist section width b_r and the equation of intersecting line between pipe internal surface and rib plate; draw an elliptic curve or a parabolic curve for rib plate internal edge passing through points A and B. The equation of the elliptic curve of rib plate internal edge (Figure A.0.2-8) is shown in Formula (A.0.2-32) and the equation of the parabolic curve of rib plate internal edge (Figure A.0.2-9) is shown in Formula (A.0.2-33). The equation of the elliptic curve is not unique and the equation of the elliptic curve with well-distributed stress and smaller stress extreme value in rib plate internal edge can be calculated by the finite element method.

$$\frac{(x-x_0)^2}{a'^2} + \frac{y^2}{b'^2} = 1 \qquad (A.0.2\text{-}32)$$

$$y^2 = \frac{y_b^2}{x_a}(x_a - x) \qquad (A.0.2\text{-}33)$$

5) Elliptic curve should be used for rib plate external edge. The distance between rib plate external edge and pipe shell shall meet the welding requirements, which should not be less than 50 mm at rib plate waist section and not be greater than 10 mm at rib plate top section.

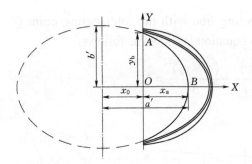

Figure A.0.2-8　Elliptic curve of rib plate internal edge

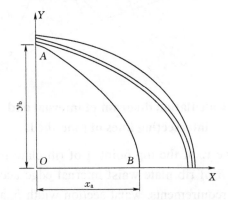

Figure A.0.2-9　Parabolic curve of rib plate internal edge

46

Appendix B Structural Analysis Methods for Underground Crescent-Rib Reinforced Bifurcations

B.1 Structural Analysis of Internal Pressure on Bifurcations

B.1.1 The gap of underground crescent-rib reinforced bifurcation can be calculated by the following formula:

$$\delta_2 = \delta_b + \delta_s + \delta_r \qquad (B.1.1\text{-}1)$$

where

δ_2 is the total gap of steel bifurcations, consisting of construction gap δ_b, cold shrinkage gap δ_s of steel bifurcations and cold shrinkage gap δ_r of surrounding rocks (mm);

δ_b is the construction gap (mm);

δ_s is the cold shrinkage gap of steel bifurcations; the gap under the lowest operation temperature δ_{s1} shall be taken for calculating the largest gap (mm);

δ_r is the cold shrinkage gap of surrounding rocks (mm).

Construction gap δ_b, steel bifurcation cold shrinkage gap δ_s and surrounding rock cold shrinkage gap δ_s can be taken or calculated as follows:

1 When concrete liner is compacted, and backfill and contact grouting is reliable, $\delta_b = 0.2$ mm.

2 Bifurcation cold shrinkage gap under the lowest operation temperature δ_{s1} can be calculated by the formula:

$$\delta_{s1} = \Delta T_s \alpha_s r_0 (1 + v_s) \qquad (B.1.1\text{-}2)$$

where

δ_{s1} is the steel bifurcation cold shrinkage gap under the lowest operation temperature (mm);

ΔT_s is the difference between the bifurcation starting temperature and the lowest operation temperature (°C), where the starting temperature is defined as the temperature when circumferential stress of pipe wall $\sigma_\theta = 0$ and $\delta_2 = \delta_b$, and may approximately take the average ground temperature if no measured data is available; the lowest operation temperature may approximately take the minimum water temperature;

v_s is the Poisson's ratio of steels;

α_s is the linear expansion coefficient of steels (1/°C);

r_0 is the inner radius of the largest common-tangent sphere of steel bifurcations (mm).

3 The cold shrinkage gap δ_r of surrounding rocks can be calculated by the following formula:

$$\delta_r = \Delta T_r \alpha_r r_5 \psi_r \qquad (B.1.1-3)$$

where

δ_r is the cold shrinkage gap of surrounding rocks (mm);

ΔT_r is the difference between the cavern surface rock starting temperature and lowest temperature, which may approximately take the difference between the average ground temperature and the average water temperature of the coldest three months if no measured data is available;

α_r is the linear expansion coefficient of surrounding rock (1/°C);

r_5 is the concrete liner outer radius, namely, tunnel excavation radius (mm);

ψ_r is the relative radius influencing coefficient of the fractured surrounding rock zone, which can be obtained from Figure B.1.1 according to r_6/r_5;

r_6 is the outer radius of the fractured surrounding rock zone, $r_6 = r_5$ for hard intact surrounding rocks, $r_6 = 7r_5$ for fractured or weak surrounding rocks, r_6 is obtained by interpolation method for medium surrounding rocks (mm).

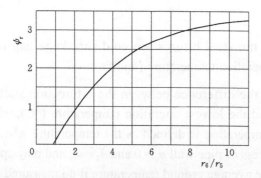

Figure B.1.1 Relation between ψ_r and r_6/r_5

B.1.2 When internal water pressure is shared by steel bifurcations and surrounding rocks, the rock cover thickness shall satisfy all the formulae below:

$$H_r \geq 6r_5 \tag{B.1.2-1}$$

$$H_r \geq \frac{p_2}{\gamma_r \cos \alpha} \tag{B.1.2-2}$$

$$p_2 = \overline{\lambda} p \tag{B.1.2-3}$$

$$\overline{\lambda} = 1000 K_0 \left(\frac{r_0}{E_{s2}} - \frac{\delta_2}{\sigma_R} \right) \frac{\sigma_R}{p r_0} \tag{B.1.2-4}$$

$$E_{s2} = \frac{E_s}{1 - v_s^2} \tag{B.1.2-5}$$

where

H_r is the thickness of overlying rocks (Figure B.1.2), where completely- and highly-weathered layers shall not be counted (mm);

p_2 is the average internal pressure shared by surrounding rocks (N/mm²);

$\overline{\lambda}$ is the average sharing ratio of surrounding rocks, which can be estimated by Formula (B.1.2-4) and be determined by the finite element method;

p is the design internal water pressure (N/mm²);

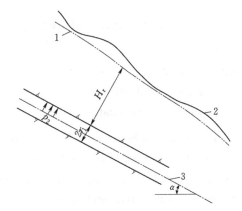

Key

1   lower limit of highly-weathered bedrock

2   ground surface line

3   pipe axis

Figure B.1.2   Thickness of the overlying rock

γ_r　is the smaller value of unit weight of surrounding rocks (N/mm³);

α　is the included angle of pipe axis to the horizontal (°), $\alpha = 60°$ when $\alpha > 60°$;

K_0　is the unit resistance coefficient of rocks (N/mm³);

E_{s2}　is the elastic modulus of steel for plane strain (N/mm²);

E_s　is the elastic modulus of steel (N/mm²);

σ_R　is the resistance limit of underground pipe membrane stress (N/mm²).

B.1.3 When internal water pressure is shared by steel bifurcations and surrounding rocks, the structural design can be conducted in the following steps:

1　Calculate the gap, taking the steel bifurcation as a cylindrical pipe with the same diameter as the common-tangent sphere.

2　Estimate the surrounding rock sharing ratio, taking the steel bifurcation as a cylindrical pipe with the same diameter as the common-tangent sphere; and estimate the internal water pressure shared by the steel bifurcation by the formula:

$$p_1 = (1 - \bar{\lambda})p \tag{B.1.3}$$

where

p_1　is the internal water pressure shared by the steel bifurcation (N/mm²).

3　Initially propose the shape and structural dimensions of the steel bifurcation as exposed steel bifurcation, according to internal water pressure shared by the steel bifurcation.

4　Reasonably simulate the gap between steel bifurcations and surrounding rocks by using 3D finite element method; conduct structural analysis of the initially proposed steel bifurcation shape by considering the joint actions of surrounding rocks and steel bifurcations.

5　Judge the rationality of initially proposed shape and structural dimensions of the steel bifurcations according to preliminary analysis results and steel resistance limit, and conduct shape optimization when necessary.

6　Analyze the sensitivity of the final shape to different gaps and surrounding rock elastic resistance coefficients.

7 Propose the design parameters and determine the shape and structural dimensions of the steel bifurcation according to sensitivity analysis results, project characteristics and engineering analogy.

8 Check the proposed design parameters, shape and structural dimensions of steel bifurcation against the exposed bifurcation criterion.

B.2 Calculation of Critical External Buckling Pressure of Steel Bifurcations

B.2.1 The external buckling pressure of the steel bifurcation can be estimated taking the steel bifurcation as a cylindrical pipe with the same diameter as the common-tangent sphere.

B.2.2 The critical external buckling pressure of smooth pipe can be calculated by empirical formula or the Amstutz formula, and shall meet the following requirements:

1 By the empirical formula:

$$p_{cr} = 612\left(\frac{t}{r}\right)^{1.7} R_e^{0.25} \qquad (B.2.2\text{-}1)$$

where

p_{cr} is the critical external buckling pressure (N/mm^2);

R_e is the yield strength of steels (N/mm^2), which shall take f_{sk} from Table 4.0.4 of this code;

r is the radius of common-tangent sphere (mm);

t is the wall thickness of main cone (mm).

2 By the Amstutz formula or consulting Figure B.2.2.

$$p_{cr} = \frac{\sigma_k}{\dfrac{r}{t}\left(1+0.35\dfrac{r}{t}\dfrac{R_{e2}-\sigma_k}{E_{s2}}\right)} \qquad (B.2.2\text{-}2)$$

σ_k can be obtained from Formula (B.2.2-3) by trial method:

$$\left(E_{s2}\frac{\delta_{2p}}{r}+\sigma_k\right)\left[1+12\left(\frac{r}{t}\right)^2\frac{\sigma_k}{E_{s2}}\right]^{\frac{3}{2}} = 3.46\frac{r}{t}(R_{e2}-\sigma_k)\left(1-0.45\frac{r}{t}\frac{R_{e2}-\sigma_k}{E_{s2}}\right)$$

$$(B.2.2\text{-}3)$$

$$R_{e2} = \frac{f_{sk}}{\sqrt{1-v_s+v_s^2}} \qquad (B.2.2\text{-}4)$$

$$\delta_p = \frac{p_2 r_5}{1000 K_{01}} \left(1 - \frac{E_{r0}}{E_r}\right) \qquad (B.2.2\text{-}5)$$

where

σ_k is the average stress at the buckling of pipe shell resulting from external pressure (N/mm²);

R_{e2} is the yield stress of steels for plane strains (N/mm²);

f_{sk} is the characteristic value of tensile strength of steels (N/mm²), as per Table 4.0.4;

δ_{2p} is the sum of δ_2 and surrounding rock plastic compression gap δ_p (mm), $\delta_{2p} = \delta_2 + \delta_p$;

δ_p is the surrounding rock plastic compression gap (mm);

K_{01} is the probable maximum value of surrounding rock unit resistance coefficient (N/mm³);

E_{r0} is the deformation modulus of surrounding rocks (N/mm²);

E_r is the elastic modulus of surrounding rocks (N/mm²).

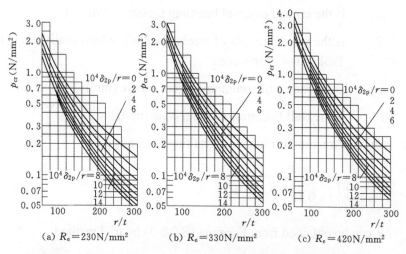

Figure B.2.2   Amstutz curves for underground pipes

B.2.3 Calculation of critical external buckling pressure of steel bifurcations with stiffener rings shall meet the following requirements:

    1   The critical external buckling pressure of pipe shell between stiffener rings shall be calculated by Mises formula:

$$p_{cr} = \frac{E_s t}{(n^2-1)\left(1+\dfrac{n^2 l^2}{\pi^2 r^2}\right)^2 r} + \frac{E_s}{12(1-v_s^2)}\left(n^2-1+\frac{2n^2-1-v_s}{1+\dfrac{n^2 l^2}{\pi^2 r^2}}\right)\frac{t^3}{r^3}$$

(B.2.3-1)

$$n = 2.74\left(\frac{r}{l}\right)^{\frac{1}{2}}\left(\frac{r}{t}\right)^{\frac{1}{4}}$$

(B.2.3-2)

where

p_{cr} is the critical external buckling pressure (N/mm^2);

n is the wave number under minimum critical pressure, rounded to the nearest integer;

l is the spacing of stiffener rings (mm).

The critical external buckling pressure can be obtained from Figure B.2.3.

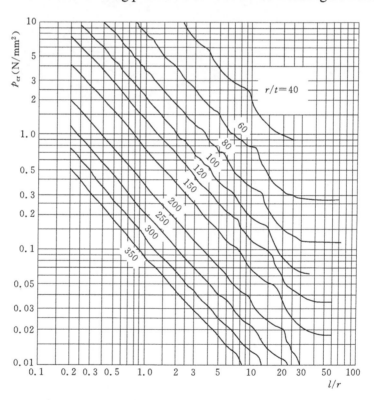

Figure B.2.3 Steel bifurcation critical external pressure curves

2 Stability analysis of stiffener rings. The critical external pressure p_c can be calculated by the formulae:

$$p_{cr} = \frac{R_e A_R}{rl}$$

(B.2.3-3)

$$A_R = ha + t\left(a + 1.56\sqrt{rt}\right) \qquad (B.2.3-4)$$

where

A_R is the effective cross-sectional area of stiffener ring (mm²);

h is the height of stiffener ring (mm);

a is the thickness of stiffener ring (mm).

3   Stiffener ring stress can be calculated by the boiler formula.

Explanation of Wording in This Code

1. Words used for different degrees of strictness are explained as follows in order to mark the differences in executing the requirements in this code.

    1) Words denoting a very strict or mandatory requirement:

        "Must" is used for affirmation; "must not" for negation.

    2) Words denoting a strict requirement under normal conditions:

        "Shall" is used for affirmation; "shall not" for negation.

    3) Words denoting a permission of a slight choice or an indication of the most suitable choice when conditions permit:

        "Should" is used for affirmation; "should not" for negation.

    4) "May" is used to express the option available, sometimes with the conditional permit.

2. "Shall meet the requirements of…" or "shall comply with…" is used in this code to indicate that it is necessary to comply with the requirements stipulated in other relative standards and codes.

List of Quoted Standards

GB/T 700,   *Carbon Structural Steels*

GB/T 713,   *Steel Plates for Boilers and Pressure Vessels*

GB/T 1591,   *High Strength Low Alloy Structural Steels*

GB/T 2970,   *Method for Ultrasonic Testing of Thicker Steel Plates*

GB/T 5313,   *Steel Plates with Through-Thickness Characteristics*

GB/T 16270, *High Strength Structural Steel Plates in the Quenched and Tempered Condition*

GB/T 19189, *Quenched and Tempered High Strength Steel Plates for Pressure Vessels*

GB/T 31946, *Steel Plates for Steel Penstock in Hydropower Station*

GB 50199,   *Unified Standard for Reliability Design of Hydraulic Engineering Structures*

GB 50766,   *Code for Manufacture Installation and Acceptance of Steel Penstocks in Hydroelectric and Hydraulic Engineering*

NB/T 35021, *Design Code for Surge Chamber of Hydropower Stations*

NB/T 35056, *Design Code for Steel Penstocks of Hydroelectric Stations*

YB/T 4137,   *Low Welding Crack Susceptibility for High Strength Steel Plates*